DR. SUSIE

SMASHES ABOUT

FROGS

Created by **Sammie Kyng**   Illustrated by **Achmad Arsad**

What does Smash Mean?

Seeking Many Answers & Sharing Honestly

For my 3 Children

CAMILET

That's a SMASH-UP of
Cameron, Dominic & Violet

Thank You SO Much for Supporting Me During My Great Transition

A Very Special Thank You to **Ceola**

She Reviewed this Smash Book and Gave Fantastic 5-Year-Old Feedback!

Hi

I'm Susie

Susie **Smash** if you please

Doctor Susie Smash

to be even **More** Pacific

I gave this name to myself

You can call me
Doctor SMASH

Today we're
Smashing
(That's Searching for Many Answers and Sharing Honestly)

about **frogs**

to be even more
Pacific!

The frogs I know live
in a river

They lay their eggs
In the water

A bunch of frog eggs
are called
a **frogspawn**

Not Song

Visit Susie and her Animal Friends at
SmashPlace.com

When the spawn hatch
they turn into
baby tadpoles

I call them **Squirmers**

Because They
Have **No** Lungs

Just Gills
Like Fish

Aren't They **Cute?**

They make me **happy**

Soon they **grow up** into

brown, green and sometimes yellow **Big** frogs

What should we call this one?
I know

How about Sebastian?
Bash for short

Because **Bash** rhymes with

You got it
SMASH

Visit Susie and her Animal Friends at
SmashPlace.com

Visit Susie and her Animal Friends at
SmashPlace.com

That's right
You need to **soak** it in

Through your **Skin**

If you dry out

You **just** might

Visit Susie and her Animal Friends at
SmashPlace.com

That's Right

Frog eyes rotate
All around

This helps them catch their **prey**

Not Hay

Prey

Prey is their **food**

They love **Bugs**

They catch **bugs**

with Really **Long** tongues

Come on **Bash**

What should we **Smash** about next?

"RiBBiT"

Visit Susie and her Animal Friends at
SmashPlace.com

We should **VISIT** a Lion?

Visit Susie and her Animal Friends at
SmashPlace.com

Lions have Amazing **hearing**

Do you hear that?
A Lion is calling us from Africa

Come on **Bash**
Let's **Smash** this

Meet Sammie Kyng

Sammie is a mother of 3 and a lover of nature and animals

She is the proud creator of Dr. Susie Smash, a young scientist - *Sammie's alter-ego* - and her best friend Bash (He's a **Frog**)

Here she is Smashing at her home in Minnesota, USA

Join Susie and Bash as they study animals, critters, bugs, furry (and slimy) things and more

Visit Susie and her Animal Friends at:

SmashPlace.Com   Instagram

Visit Susie and her Animal Friends at
SmashPlace.com

Copyright© MMXXI Sammie Kyng

All Rights Reserved

Request permission to reproduce **more than a short quote** of any part of this book at:

SmashPlace.com

Instagram.com/DrSusieSmash

ISBN: 978-1-7361341-0-8

Kyngdom Publishing

Visit Susie and her Animal Friends at
SmashPlace.com

Did you Love
Smashing about Frogs
with Susie and Bash?

Please Share Your Review

Visit Susie and her Animal Friends at
SmashPlace.com

Do you know who the **Best Hunter** is in a Pride of Lions?

Let's Smash about it with Dr. Susie and Bash!

SmashPlace.com

Visit Susie and her Animal Friends at
SmashPlace.com

Made in the USA
Las Vegas, NV
10 December 2021

36821012R00026